CONCEPTS AND TECHNIQUES IN MODERN GEOGRAPHY No. 6

CLASSIFICATION IN GEOGRAPHY

by

R.J. Johnston

(University of Sheffield)

CONTENTS

Acknowledgement

I am very grateful to Stan Gregory, Alan Hay and Peter Taylor for their
helpful comments on various drafts of this manuscript.

The permission of the following for the reproduction of copyright material
is gratefully acknowledged:

Dr L. Adrian and University of Wales Press for figure 15.
Dr F.L. Jones for figure 10 and table 5.
J. Byfuglien, A. Nordgard and Universitets forlaget (Oslo)
for figures 12 and 13.

CLASSIFICATION IN GEOGRAPHY

I INTRODUCTION

The aim of any scientific activity is to understand; once we can explain why certain events occur (and when, and where) then we might be in a position to control them, so as to produce a better world. Thus science involves the development of predictive methodologies, which produce generalisations, theories, and laws. These apply not to particular instances or events but to classes of phenomena; we develop theories and laws not for one individual pro-glacial stream, but for all such streams. In other words, we define a category whose members we call pro-glacial streams; we accept that each of these streams has characteristics peculiar to itself - its location, for example - but argue that it has sufficient characteristics in common with all other pro-glacial streams that we can study them as a group.

Adoption of this scientific philosophy immediately raises the question of category definition: what are the individuals that we can group together to form a particular phenomenon class? A logical, or deductive, model of our subject matter may define our classes for us, as is the case with the set of rules that geomorphologists have developed for stream ordering: Grigg (1965) gives a general introduction to this mode of investigation. But what if we have no predetermined set of rules, and have to work in an inductive fashion, delimiting categories among the phenomena and/or objects that we have observed? With streams, for example, we may wish to study them not only according to their order in the Strahler-ordering system, but also by their discharge features, their thalweg profiles, and so on. For these last variables we probably have no predetermined categories - and if we have, they are probably based on arbitrary boundary lines. And so we must find a means of grouping the streams together according to the variables we decide to use. This is the classification process, by which individual objects are allocated, as objectively as possible, to discrete categories.

The philosophy of science is based on an acceptance of the classification axiom, that there are groups of like phenomena/objects which can be treated as a single unit for the purpose of making valid generalisations about aspects of their behaviour. Clearly, many different classifications are possible, depending on the purpose of the study being undertaken: streams, for example, form a single class, but they can be subdivided on a variety of criteria, such as order, depth, gradient, and so on. The purpose of classification procedures is to provide a grouping which is valid for the scientific activity being undertaken: criteria for classification must first be determined, and then the objects measured on these must be allocated to classes. The first part of this procedure lies in the substantive field; the second part - the choice and operation of a classification process - is the focus of this monograph.

There is a long history of classification work in geography, of which the best examples probably come from climatology. The definition of climatic regions has been undertaken by many scholars. Some of their efforts have been preliminaries to predictive work on, for example, the relationships between climate and crop yields. Many others have been undertaken for didactic purposes only, as means of generalising about the complex map of climates on the ground, for which vast volumes of material are available. None of us is able to assimilate the mass of material available in daily climatic records, the plethora of census volumes, or the myriad aerial and space photographs which are filling our libraries at increasing rates. We may just sample from the relevant information sets, or we may try and reduce the 'information overload', just as we reduce, say, the figures on the percentage of the workforce in each English county into a few percentage classes, enabling the construction of a simple choropleth map which indicates the salient features of the national pattern.

There are two reasons for classification, therefore. The first involves a scaling factor, reducing a large number of individuals to a small number of groups, to facilitate description and illustration. Within geography, the large literature on the functional classification of towns (Smith, 1966) is an example of such work. The second comprises the definition of phenomena-classes about which general statements are to be derived, as is the case in the work on central places, in which classes of towns are defined as a prelude to studies of such topics as town spacing and consumer behaviour. Whichever of these two purposes is relevant, the methods are the same, since the principles are identical: the combination of discrete individuals with similar individuals to produce groups. Many alternative sets of rules are available in the application of these methods. Choice between them depends on several factors, such as the nature of the measurement for the objects being studied and the goals of the classification. Two main types ('a classification of classification procedures') exist, however, and these - termed agglomerative and divisive methods - are introduced in the following sections.

II AGGLOMERATIVE METHODS

These methods start with a lot of individuals and proceed, using a set of predetermined rules, to allocate them to groups according to levels of similarity on the chosen criteria. Perhaps the ideal procedure would be for the classifier to produce every grouping possible, and then to decide which is the best for his particular purpose. But this is almost always an impossibility, because of the large number of individuals involved. For example, if four individuals are to be allocated to one of two groups, there are seven possible combinations, as follows:

Combination	Group 1	Group 2
A	1,2	3,4
B	1,3	2,4
C	1,4	2,3
D	1	2,3,4
E	2	1,3,4
F	3	1,2,4
G	4	1,2,3

If there are five individuals, there are 15 possibles (work out what they are), and so on. And since classification is almost certainly involved with large numbers of individuals, searching the total set of possibilities for the optimum is clearly an impossibility (even with a high speed computer). Thus researchers interested in classification, in a wide range of disciplines, have been forced to develop methods (often termed methods in numerical taxonomy), which will produce the 'best' classification for their particular data set. It is with some of these methods - in general, the simpler - that we are concerned here.

(i) Elementary Linkage Analysis

We have data on two variables for sixteen villages lying to the northwest

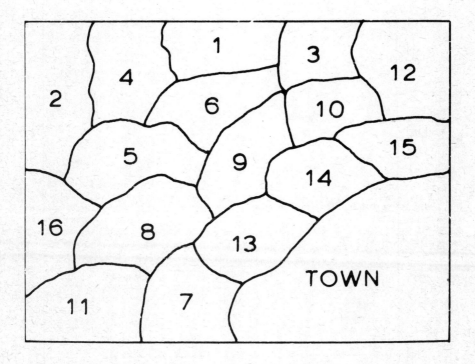

Fig. 1 A hypothetical set of sixteen villages.

of a large town (Fig. 1), the two variables being the percentage of the village workforce who commute to the town for employment (variable X) and the percentage of the village population who were born within the local are a (variable Y). A two-dimensional scattergram (Fig. 2) shows the values for each village on these two variables; our task is to classify the villages into groups which contain members that are similar in their position on the two axes.

The first task is to produce a measure of the degree of similarity between each pair of villages. This is done by using the well known Pythagoras' Theorem from Euclidean geometry, which states that the square of the distance

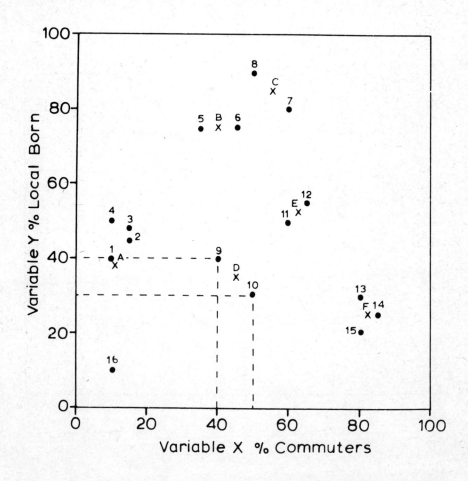

Fig. 2 The villages of Figure 1 according to their values on two variables.

on the hypotenuse of a right-angled triangle equals the sum of the squares of the distances on the other two sides (Fig. 3). Thus in general terms:

$$D_{ij} = \sqrt{(i_x - j_x)^2 + (i_y - j_y)^2} \quad \dots\dots\dots\dots\dots\dots\dots(1)$$

where i_x, j_x = the values for i and j on variable x

$\qquad\quad i_y, j_y$ = the values for i and j on variable y

and $\quad D_{ij}$ = the distance between i and j

Figure 2 shows how we apply equation (1) to finding the distance between places 9(i) and 10(j). The term $(i_x - j_x)^2$ is $(40 - 50)^2$ or 100. Similarly $(i_y - j_y)^2$ is $(40 - 30)^2$ or 100, so $D_{ij} = \sqrt{100 + 100}$, which is 14.1.

In turn, i and j represent each of sixteen villages, so that we finish up with the 16 x 16 matrix of inter-village 'distances' that is given as Table 1. The values on the principal diagonal of this matrix (from top left to bottom right) are zero, indicating that village i is completely similar to village j, when i = j; the values to the right of the principal diagonal are the mirror image of those to the left, which means that we have a square symmetric matrix.

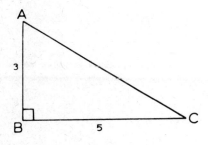

PYTHAGORAS' THEOREM

$$AC^2 = AB^2 + BC^2$$
$$= 3^2 + 5^2$$
$$= 9 + 25 = 34$$
$$\therefore AC = \sqrt{34} \quad = 5\cdot831$$

Fig. 3 Pythagoras' Theorem.

7

We can now proceed to the classification, using the criterion that each place is grouped along with that to which it is most similar. (This will be its nearest neighbour on our scattergram in Fig. 2). The nearest neighbour is identified from the distance matrix by finding, for each column, the row with the smallest value in it - excluding the zero values, along the principal diagonal, which indicate distance to self. Thus for column 1 in Table 1, the smallest value is 7.1 in row 2; 1's nearest neighbour is 2. The total set of nearest neighbours is:

Column	1	2	3	4	5	6	7	8
Smallest row value is in row	2	3	2	3	6	5	8	7

Column	9	10	11	12	13	14	15	16
Smallest row value is in row	10	9	12	11	14	13 15	14	1

(Note that place 14 has two equidistant nearest neighbours)

The groups can be shown by a diagram, in which an arrow links each place to its nearest neighbour, as follows:

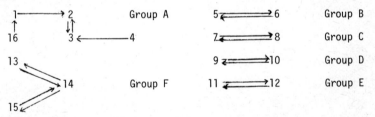

The sixteen villages have been classified into six groups; the similarity between the pairs of villages is their distance on the Cartesian coordinates of the graph (Fig. 2), and the classification criterion allocates them to groups with their nearest neighbours, those they are most similar to.

(ii) Grouping the Groups

But we needn't stop here; we can proceed to 'group the groups'. To do this, we first produce a matrix of the distances between the groups (Table 2). Each of the off-diagonal distances is the average distance between members of the two pairs: thus the distance between groups B and C is calculated as:

Table 1 Distances between places on Figure 2.

								Place								
	1	2	3	4	5	6	7	8	9	10	11	12	13	14	15	16
1	0	7.1	9.4	10.0	43.0	49.5	64.0	64.0	30.0	41.2	51.0	57.0	70.0	76.5	72.8	30.0
2	7.1	0	3.0	7.1	36.1	42.4	57.0	57.0	25.5	38.1	45.3	51.0	66.7	72.8	69.6	35.4
3	9.4	3.0	0	5.4	33.6	40.4	55.2	54.7	26.3	39.4	45.1	50.5	67.5	73.7	70.8	38.3
4	10.0	7.1	5.4	0	35.4	43.0	58.3	56.6	31.6	44.7	50.0	55.2	72.8	79.1	76.2	40.0
5	43.0	36.1	33.6	35.4	0	10.0	25.5	21.2	35.4	47.4	35.4	36.1	63.6	70.7	71.1	69.6
6	49.5	42.4	40.4	43.0	10.0	0	15.8	15.8	35.4	45.3	29.2	28.3	57.0	64.0	65.2	73.8
7	64.0	57.0	55.2	58.3	25.5	15.8	0	14.1	44.7	51.0	30.0	25.5	53.9	62.1	63.3	86.0
8	64.0	57.0	54.7	56.6	21.2	15.8	14.1	0	51.0	60.0	41.2	38.1	67.1	73.8	76.2	89.4
9	30.0	25.5	26.3	31.6	35.4	35.4	44.7	51.0	0	14.1	22.4	29.2	41.2	46.4	44.7	42.4
10	41.2	38.1	39.4	44.7	47.4	45.3	51.0	60.0	14.1	0	22.4	29.2	30.0	35.4	31.6	44.7
11	51.0	45.3	45.1	50.0	35.4	29.2	30.0	41.2	22.4	22.4	0	7.1	28.3	35.4	36.1	64.0
12	57.0	51.0	50.5	55.2	36.1	28.3	25.5	38.1	29.2	29.2	7.1	0	29.2	36.1	38.1	71.1
13	70.0	66.7	67.5	72.8	63.6	57.0	53.9	67.1	41.2	30.0	28.3	29.2	0	7.1	10.0	72.8
14	76.5	72.8	73.7	79.1	70.7	64.0	62.1	73.8	46.4	35.4	35.4	36.1	7.1	0	7.1	76.5
15	72.8	69.6	70.8	76.2	71.1	65.2	63.3	76.2	44.7	31.6	36.1	38.1	10.0	7.1	0	70.7
16	30.0	35.4	38.3	40.0	69.6	73.8	86.0	89.4	42.4	44.7	64.0	71.1	72.8	76.5	70.7	0

Pairs of Places	Distance
5 - 7	25.5
5 - 8	21.2
6 - 7	15.8
6 - 8	15.8

Σ78.3 average = 78.3/4 = 19.6

These off-diagonal distances are, in fact, the distances between the group centroids, which are the central foci of the groups (shown by Xs in Fig. 2). In a two-place group, the centroid is midway between the members; in groups with more than two members, the centroid is at the point from which distances to all members sum to the smallest possible value. (For the moment, we will ignore the values on the principal diagonal of Table 2).

Table 2 Inter-group distances from Table 1[a]

(members)	A (1-2-3-4-16)	B (5-6)	C (7-8)	D (9-10)	E (11-12)	F (13-14-15)
A	18.6	46.7	64.2	36.4	54.0	72.6
B	46.7	10.0	19.6	40.9	32.2	65.3
C	64.2	19.6	14.1	51.7	33.7	66.1
D	36.4	40.9	51.7	14.1	25.8	38.2
E	54.3	32.2	33.7	25.8	7.1	33.9
F	72.0	65.3	66.1	38.2	33.9	8.1

[a]values along the principal diagonal (in boxes) show the average intra-group distances

Grouping from Table 2 proceeds in exactly the same way as the initial grouping from Table 1. First we find the smallest row value in each column, as follows:

Column	A	B	C	D	E	F
Smallest row value in row	D	C	B	E	D	D

From this we derive our grouping:

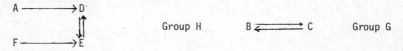

A ──────→ D

F ──────→ E

Group H   B ⇄ C   Group G

10

And so we can go on, producing the distance matrix in Table 3, and grouping the groups so that all sixteen villages are now part of one large group.

Table 3 Inter-group distances from Table 2[a]

| | Group | |
	G (B-C)	H (A-D-E-F)
G (B-C)	17.1	52.7
H (A-D-E-F)	52.7	37.2

[a] values along the principal diagonal (in boxes) show the average intra-group distances

We have proceeded then, from sixteen villages through six groups and then two groups, until every village is in the same group. This process can be represented by a linkage tree (Fig. 4), in which the linkages at each step are indicated. (Note that the villages are arranged in a specific order along the horizontal axis to avoid lines crossing).

Two problems arise from this set of procedures. The first is: 'when do we stop it?'. Clearly to proceed until all of the villages are in one group is worthless, so when should the procedure be halted? We may have decided beforehand that we want six groups, so we would stop after the first step,

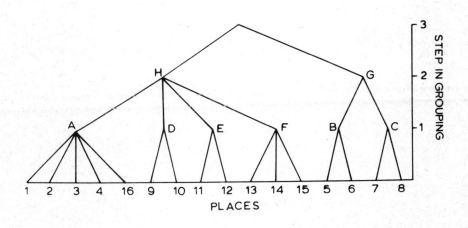

Fig. 4 Linkage tree for Elementary Linkage Analysis of Villages in Figure 2.

11

but if we had no such prior decision, we might need a guideline. One such is the average intra-group distance, shown in the principal diagonals of the matrices of Tables 2 and 3. These are the average distances among members of the particular group, so that in Table 3 the value of 17.1 for group C is calculated as:

Pairs of Places	5-6	5-7	5-8	6-7	6-8	7-8	
Distance	10.0	25.5	21.2	15.8	15.8	14.1	Σ= 102.4
	average distance = 102.4/6 = 17.1						

We may decide that the groups used in Table 3 are too large, in the sense that the average distance within them is very great, so that the members are really not very similar. In general, the intra-group distances in Table 2 are smaller, so we may decide to stick with the six-group solution.

The second problem is that the procedure does not allow us much choice. We can have sixteen unique villages, six groups of villages, two groups of villages, or all sixteen in one group. What if we want four groups? Further, some of the groups in Table 2 have low intra-group average distances, suggesting very cohesive groups of similar villages, whereas others have much larger distances. It would be preferable to be able to select the first type, and do without the others.

(iii) Hierarchical Procedures

Because of the problems just discussed, the method outlined above is usually adopted in a modified sense, with the grouping procedure moving more slowly. It operates under the same rules, except that at each step, only the pair with the shortest distance between them are grouped, rather than all pairs being grouped at once. This gives greater flexibility in deciding where to stop the process - or which grouping to select after the whole process has been completed and the linkage tree drawn.

This hierarchical procedure starts with Table 1. For each column of the matrix, the row with the smallest entry is noted (excluding the principal diagonal, as usual), as follows:

Column	1	2	3	4	5	6	7	8
Row with smallest entry	2	3	2	3	6	5	8	7
value	7.1	3.0	3.0	5.4	10.0	10.0	14.1	14.1

Column	9	10	11	12	13	14	15	16
row with smallest entry	10	9	12	11	14	13/15	14	1
value	14.1	14.1	7.1	7.1	7.1	7.1	7.1	30.0

12

The smallest of these values is then isolated, and is used to determine which pair will be grouped. The distance is 3.0 units, between villages 2 and 3, and these form group A at the first step in the hierarchical procedure.

We must now produce a new distance matrix, in which the separate villages 2 and 3 are replaced by group A (Table 4). This is a 15 x 15 matrix, in which the distances in column A and row A are the averages for all members of group A to the relevant village (i.e. the distance from the latter to the centroid for group A); these are computed in exactly the same way as before, so that the distance from village 1 to group A is the mean of the distances between 1 and 2 and 1 and 3 (7.1 and 9.4, making 8.3). Note that we now have an intra-group mean distance in the principal diagonal at the row A/column A inter-section of Table 4.

Having computed the distance matrix, we now search it for the shortest distance, which will indicate the pair of villages to be grouped at the next step. The distance is 6.2, and groups village 4 with group A. Again, a new distance matrix is computed, this time 14 x 14 (it is not given here, because of space considerations), and we proceed to the next step. The process con-tinues, one grouping at a time, until all villages are in the one group. Its parameters are summarised as follows:

Step	Places Grouped	Distance	Forming Group	Average Intra-Group Distance
1	2-3	3.0	A	3.0
2	A-4	6.2	B	5.2
3	11-12	7.1	C	7.1
	13-14-15	7.1	D	8.1
4	B-1	8.8	E	7.0
5	5-6	10.0	F	10.0
6	7-8	14.1	G	14.1
	9-10	14.1	H	14.1
7	F-G	19.6	J	17.1
8	C-H	25.8	K	20.7
9	D-K	36.1	L	27.7
10	E-16	37.9	M	18.6
11	L-J	50.8	N	38.3
12	N-M	68.7	O	30.1

The linkage tree for this grouping is Figure 5, with the distance involved in the grouping forming the vertical axis. From this, we get a clear impres-sion of the short distances involved in the first few groups (up to J, say), and the larger ones thereafter. If we have no a priori reason for accepting a certain number of groups, this diagram should help us decide where the group-ing should be stopped. One important difference to note between this grouping

Table 4 Distances between places after first step in hierarchical grouping

						Place								
	1	4	5	6	7	8	9	10	11	12	13	14	15	16	A
1	0	10.0	43.0	49.5	64.0	64.0	30.0	41.2	51.0	57.0	70.7	76.5	72.8	30.0	8.3
4	10.0	0	35.4	43.0	58.3	56.6	31.6	44.7	50.0	55.2	72.8	79.1	76.2	40.0	6.2
5	43.0	35.4	0	10.0	25.5	21.2	35.4	47.4	35.4	36.1	63.6	70.0	71.1	69.6	34.8
6	49.5	43.0	10.0	0	15.8	15.8	35.4	45.3	29.2	28.3	57.0	64.0	65.2	73.8	41.1
7	64.0	58.3	25.5	15.8	0	14.1	44.7	51.0	30.0	25.5	53.9	62.1	63.3	86.0	56.1
8	64.0	56.6	21.2	15.8	14.1	0	51.0	60.0	41.2	38.1	67.1	73.8	76.2	89.4	55.6
9	30.0	31.6	35.4	35.4	44.7	51.0	0	14.1	22.4	29.2	41.2	46.4	44.7	42.4	25.9
10	41.2	44.7	47.4	45.3	51.0	60.0	14.1	0	22.4	29.2	30.0	35.4	31.6	44.7	38.8
11	51.0	50.0	35.4	29.2	30.0	41.2	22.4	22.4	0	7.1	28.3	35.4	36.1	64.0	45.2
12	57.0	55.2	36.1	28.3	25.5	38.1	29.2	29.2	7.1	0	29.2	36.1	38.1	71.1	50.8
13	70.7	72.8	63.6	57.0	53.9	67.1	41.2	30.0	28.3	29.2	0	7.1	10.0	72.8	67.1
14	76.5	79.1	70.0	64.0	62.1	73.8	46.4	35.4	35.4	36.1	7.1	0	7.1	76.5	73.3
15	72.8	76.2	71.1	65.2	63.3	76.2	44.7	31.6	36.1	38.1	10.0	7.1	0	70.7	70.2
16	30.0	40.0	69.6	73.8	86.0	89.4	42.4	44.7	64.0	71.1	72.8	76.5	70.7	0	32.7
A	8.3	6.2	34.8	41.1	56.1	55.6	25.9	38.8	45.2	50.8	67.1	73.3	70.2	32.7	3.0

and that produced by Elementary Linkage Analysis for the same set of villages is that village 16 is not grouped with villages 1, 2, 3 and 4 until step 10; in the earlier classification, the five villages were grouped at the first step, despite the considerable distance separating 16 from the rest of the group. Our present method (Hierarchical Clustering with Centroid Replacement) leaves outliers such as village 16 as separate groups; Elementary Linkage Analysis immediately allocates an individual to the group containing its nearest neighbour, irrespective of the distance involved.

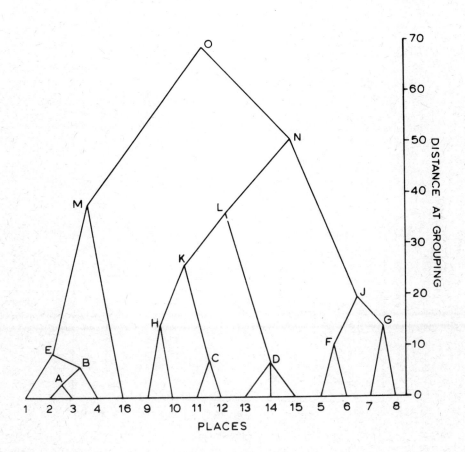

Fig. 5 Linkage tree for Hierarchical Clustering with Centroid Replacement of Villages in Figure 2.

Where should we stop the grouping process? Which step of the procedure provides the 'best' set of groups? Unfortunately, there is no simple definition of 'best', so we might have to make an arbitrary decision based on inspection of the linkage tree. Alternatively, we may say that no group will be accepted if the distance involved in its production is more than a predetermined one. In our case, this may be 20.0 distance units, in which case we would stop at step 7, with six resulting groups (E-16-H-C-D-J). Or we may say that no group will be accepted if the average intra-group distance exceeds a certain figure. (Note that the intra-group distances do not, like the grouping distances, get larger at every succeeding step of our process; the reason for this is that adding one individual to a group will probably have less influence on the size of the average intra-group distance than will adding one group to another. So if we adopted a maximum distance of 20.0 in our example, we would accept all those groups produced up to step 7, plus that from step 10). But how are such arbitrary distances to be determined, unless we have good a priori reasons for a selection? If we graph all of the distances in Table 1 (Fig. 6), there is some suggestion of a break around the distance 20.0, but the cumulative frequency distribution suggests no such break. In this, as in several other aspects of the classification procedure, arbitrary choice is necessary, and complete 'objectivity' impossible (Johnston, 1968).

A variety of other methods has been suggested. One could, for example, graph the average intra-group distance - for all groups - against the average inter-group distance at all steps of the procedure. But in the end, an arbitrary decision must be made.

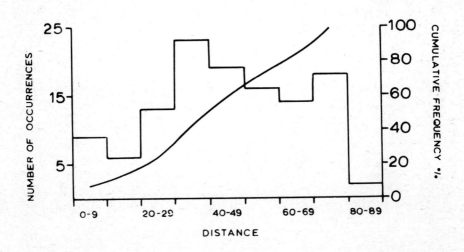

Fig. 6 Frequency Histogram and Cumulative Frequency Distribution of Distances in Table 1.

(iv) Alternative Group Definitions

The rule adopted in our hierarchical procedure was that at each step the places grouped together were replaced by their centroid, and the new distance matrix showed the distances between group centroids. (One-member groups, of course, have the centroid as the individual). But, as several writers have shown (e.g. Lance and Williams, 1967), other rules can be suggested, and logically defended on the basis of certain definitions of groups. These include:

(1) The furthest distance method, in which the distance between two groups is expressed as the maximum distance between a member of one group and a member of the other. This would emphasize group differences and, given an arbitrary stopping rule of a certain distance, undoubtedly mean more groups than would application of centroid replacement to the same data.

(2) The nearest distance method, which takes a 'liberal' definition of inter-group distance as the shortest distance between any pair of members.

(3) The group average method, in which, when multi-member groups are involved, the amalgamation of two groups results in their replacement by the centroid of their centroids, rather than by the centroid for all group members. Thus if the distance from group X (comprising places i and j) to group Y was 10.0 and from group Z (comprising places k, l, m, n) to group Y was 15.0, the distance from new group X-Z to Y would be 12.5 (the mean of 10.0 and 15.0); the larger membership of Z is irrelevant here, since the group is considered as one individual.

(4) The total distance method (sometimes known as Rank Order Typal Analysis, see Johnston, 1968), in which an individual is only allocated to a group if it is closer to all group members than it is to any other individual. Clearly, this is a very stringent criterion, a 'conservative' definition.

(v) Ward's Method

Choosing between these variants of the hierarchical procedure - and only a small sample of the possibilities has been listed here - is very difficult unless you have a clear definition of a group. And so far, we have looked only at those methods which are based on inter- and intra-group distances. One other method, increasingly popular with geographers, has been developed by Ward (1963).

Ward's method is based on an argument that the aim of classification should be to produce groups in which the distances of individual members to group centroids are kept to a minimum. In other words, the variance of the distance is to be minimised. This variance (Ward terms it the Error Sum of Squares - ESS) is:

$$(\sum_{i=1}^{n} (D_{ix})^2)/n = ESS \dots\dots\dots\dots\dots(2)$$

where D is the distance between place i and the group centroid x, assuming that i is a member of the group (X), and n in the number of members of group X

This formula can be written as:

$$\sum_{i=1}^{n} i_x^2 - ((\Sigma i_x)^2/n) = ESS$$

where i_x is the value of place i on variable x

Summation is over all variables.

Figure 7 provides an example for this method. We have ten constituencies arrayed according to the percentage of their electors who voted Labour, and we want to group constituencies. Which pair will provide the smallest value of ESS? Clearly we need consider adjacent pairs only since, for example, 1 will not group with 3 before being grouped with 2. For constituencies 1 and 2, the ESS is computed as follows:

$$ESS_{12} = (1_x^2 + 2_x^2) - ((1_x + 2_x)^2/n)$$

where 1_x = the value for constituency 1 on variable x, etc.

$$= (5^2 + 15^2) - (5 + 15)^2/n$$

$$= 250 - 400/2 = 50$$

(Note that this value - 50 - is the sum of the squared distances from places 1 and 2 to a centroid for a grouping of them, which would have a value of 10 on the percentage voting Labour variable).

For all possible pairs, we would get:

Pair	1-2	2-3	3-4	4-5	5-6	6-7	7-8	8-9	9-10
ESS	50	112.5	8	12.5	128	12.5	60.5	288	388

The minimum ESS is 8, and we group constituencies 3 and 4. At the next step, we replace 3 and 4 by group A and must work out the ESS for new possible groups:

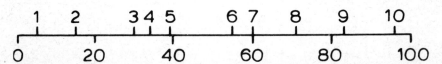

Percentage of voters in Constituency voting Labour

Fig. 7 Percentage voting Labour in ten Constituencies.

18

ESS   A - 2 (i.e. 2-3-4) = $(15^2 + 30^2 + 34^2)$ - $(79^2)/3$ = 200.7

ESS   A - 5 (i.e. 3-4-5) = $(30^2 + 34^2 + 39^2)$ - $(103^2)/3$ = 40.7

The minimum ESS is now 12.5, which leads to the grouping of constituencies 6 and 7 (the value of 12.5 for 4 and 5 is now redundant, because 4 has been grouped with 3). The full procedure for grouping our 10 constituencies is as follows:

Step	Group	ESS	Forming Group	New ESS for possible groups			
1	3-4	8	A	A-2	200.7	A-5	40.7
2	6-7	12.5	B	B-5	240.7	B-8	134.0
3	A-5	40.7	C	C-2	321.0	B-C	696.2
4	1-2	50.0	D	C-D	801.2		
5	B-8	134.0	E	E-C	1322.8	E-9	464.8
6	9-10	338.0	F	E-F	1126.0		
7	C-D	801.2	G	G-E	3457.9		
8	E-F	1126.0	H	G-H	7783.6		
9	G-H	7783.6	J				

Figure 8 is the linkage tree for this process. Since the grouping is based on mean squared distances, the great intra-group variation at the later steps is magnified (this is an advantage over the methods based on the distances themselves), and, in this case at least, the decision where to stop the grouping is fairly obvious.

(vi) Once a Group ...

Whichever of the hierarchical procedures we adopt, one problem always emerges. The procedures always group groups, without enquiring whether, at certain steps, the criteria for the classification would be better met if one of the existing groups were dismantled. For example, we might have divided Europe, for the organisation of super-soccer leagues, into three groups - Western, Central, and Eastern. At a later date, we may decide we need only two leagues. The best allocation procedure to adopt would probably be to allocate the western part of the Central League to the Western League, and the remainder to the Eastern. But if we were constrained to dealing with the Central League as a whole, the balance of the two new leagues would be uneven.

An example of this problem in grouping procedures is given by Figure 9. We have six stations located along a railway line, which we want to group into 'maintenance districts' which will minimise inter-station movements of equipment. Using hierarchical grouping with centroid replacement, the grouping would proceed as follows:

19

Step	Places Grouped	Distance	Forming Group	Average Intra-Group Distance
1	1-2	10.0	A	10.0
2	3-4	20.0	B	20.0
3	5-6	26.0	C	26.0
4	A-C	40.5	D	33.0
5	B-D	65.8	E	

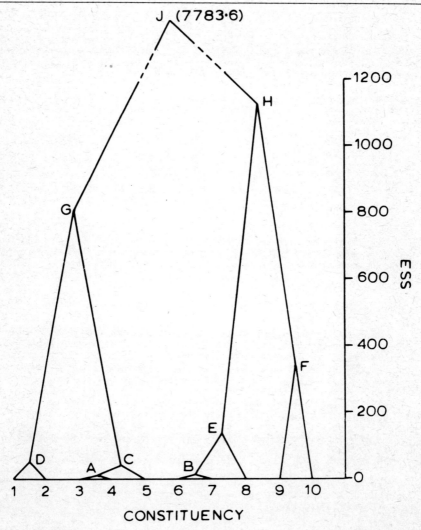

Fig. 8 Linkage tree for application of Ward's Method to Constituencies in Figure 7

If we decided on three groups each (A, B, and C) would contain two stations. But what if we decided on only two groups - perhaps in a later rationalisation programme? One would be group B, with two stations, and the other group D, with four stations, of which the most distant were 81 kilometres apart. Intuitively, we would feel that a better two-group solution would allocate station 5 to group B and station 6 to group A, which would give average intra-station distances of 28.3 and 21.7 kilometres, instead of the 20.0 and 33.0 produced by the grouping procedure.

This problem comes about because after step 3, we are grouping groups, not individual stations. We can get round it by checking whether the grouping is optimal either at every step of the grouping or, more realistically, when we feel we have the required number of groups. This involves locating the group centroids and seeing whether each individual is closest to its own group centroid. For our six stations, this produces the following distances:

Distance to Centroid of Group	Three-Group Solution			Two-Group Solution	
	A	B	C	B	D
Station					
1	5	91	45.5	91	25.25
2	5	81	35.5	81	15.25
3	75	10	35.5	10	55.75
4	95	10	55.5	10	75.75
5	52.5	32.5	13	32.5	33.25
6	27.5	57.5	13	58.5	7.25

The underlined distances indicate which group a station is allocated to.

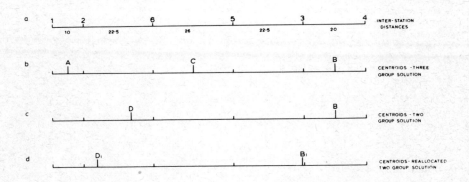

Fig. 9 The Optimal Grouping Problem: A Hypothetical Railway System

In the three-group solution, every station is closer to the centroid of the group to which it has been allocated than it is to the centroid of any other group. (We discover this by seeing whether the underlined distance is the shortest in the relevant row). But in the two-group solution, Station 5 is closer to the centroid of group B than it is to the centroid of its own group, D. It would appear to be mis-classified, so we reallocate it to group B (now termed group B^1), and rework the distances:

Distance to Centroid of Group	B^1	D
Station 1	80.2	14.2
2	70.2	4.2
3	0.8	66.8
4	20.8	86.8
5	21.7	44.3
6	47.7	18.3

There are no mis-classifications now, so the solution is deemed optimal and we finish with two groups, one containing stations 1, 2 and 6, and the other stations 3, 4 and 5. (It is of interest to note that if we had used the nearest distance, rather than the centroid replacement method, we would have achieved an optimal two-group solution, but not an optimal three-group solution: I leave you to demonstrate this).

(vii)   Into n Dimensions

All of the examples used thus far have been concerned with classification of either one-dimensional (stations along a railway line; constituencies according to the Labour vote) or two-dimensional (villages characterised by two variables) 'spaces'. This has been because 'maps' of where the places and group centroids are can be accurately reproduced in our diagrams, which makes it possible to see how the grouping procedures operate. But all of the methods we have discussed apply equally as well to n-dimensional classifications, where n ≥ 3 and our individuals are categorised on a large number of different variables.

In the one-dimensional case, distances are additive, so that where distance AB > distance AC then AC = AB + BC. In the two-dimensional case we apply the Pythagorean formula of equation (1), and this can be extended into n dimensions, as follows:

$$D_{ij} = \sqrt{(i_1 - j_1)^2 + (i_2 + j_2)^2 \ldots + (i_n - j_n)^2} \ldots \ldots (4)$$

where i_n, j_n = the values of i and j on the nth variable.

Each term within parentheses represents the distance between the two places (i and j) on the particular variable; in equation (4) we obtain the sum of the

squared distance on all variables and then take the square root of this value
to give the length of the hypotenuse in the n-dimensional space.

Problems can arise in the application of equation (4) because of scale
variations between variables. In our villages example, virtually the full
range of values from 0-100 per cent was observed for each variable. If, however, the range for one had been 0-100 and for the other only 40-60, then
clearly the former variable would contribute much more to the distance between
pairs of places than would the latter, because its range of values was five
times greater. If several variables are chosen for a classification, pre-
sumably each should have equal weight in determining the groups (though see
Johnston, 1965, 1970 for discussions of this). To ensure this, values on the
original variables are usually expressed in Z-deviate form, where:

$$Z_i = \frac{(x_i - \bar{x})}{S_x} \dots\dots\dots\dots\dots\dots\dots\dots\dots\dots\dots (5)$$

where x_i = value of variable x for ith observation

$\bar{x}$ = the mean of all values of x

S_x = the standard deviation for x

Z_i = the Z score for observation i

After this transformation, each variable has a mean of 0 and a standard deviation of 1.0, and will have the same weight as all others in the classification.

One other probable problem with multi-dimensional classification concerns
the inter-relationships among the variables. Computation of distances using
the Pythagorean formula assumes that the variables are all orthogonal, or un-
correlated: if they are not the distances are biased. To circumvent this problem, it is possible to standardise the distances, but a more usual procedure
is to replace the original set of variables by a new set, through the use of
principal components analysis. (Principal components analysis is outlined in
another of the monographs in this series - by S. Daultrey. Basically what it
does is to take a matrix of observations on a set of variables and replace the
latter by a new set of variables - usually, though not necessarily, a smaller
number - which are orthogonal. In addition, the scores for the observations
on these new variables - the components - have a mean of 0 and standard deviation of 1.0, so that Pythagorean distances can be computed immediately.
A further advantage of principal components analysis is that it removes re-
dundancies in the original data matrix. Two variables may be repetitions of
each other - average income and average years of schooling as indices of socio-
economic status, for example - and would give their joint concept undue weight
if grouping used both. Principal components analysis replaces them by a single
variable).

Once the distance matrix has been computed, grouping proceeds as before,
using whichever of the procedures is deemed apt for the particular purpose.
A linkage tree can be produced to display the grouping procedure, but unfortunately the positions of the observations in more than three-dimensional space
can not be graphically portrayed. The test for optimality of grouping can be
made; when dealing in more than one dimension this uses the technique of mul-

tiple linear discriminant analysis; the details of this method can not be outlined here (see King, 1969), but its aims and results are the same as those
discussed above in the railway station problem.

Not all classifications are based on distance matrices; a wide range of
measures of similarity and dissimilarity between observations is available
(Everitt, 1974; Spence and Taylor, 1970). One frequently used is the correlation coefficient (Pearson's, Spearman's or Kendall's) which shows the degree
of agreement between two variables over a set of observations, or of two observations over a set of variables. Hierarchical grouping procedures can operate on correlation matrices, except of course that grouping would be of the
two places with the largest value in the matrix (i.e. the most similar) rather
than the smallest. There are some problems, however, of computing average
correlations between and among groups.

(viii)   An Example: Social Areas in Melbourne

Urban geographers and sociologists have produced a voluminous literature
on social areas within cities (Johnston, 1971). A frequent aim of such work
has been the provision of a map of social areas (types of residential districts), defined according to population, household, and housing characteristics. The initial theoretical statement presented a deductive model which involved a classification scheme, but most recent work has relied on inductive
procedures, using methods such as those outlined here to suggest social area
types (Berry and Rees, 1969).

One example of such work is a study of Melbourne, Australia by Jones
(1969). He reduced 24 original variables to three orthogonal components, and
then proceeded to classify the 611 census districts of the metropolitan area,
using the hierarchical grouping procedure - based on a Pythagorean distance
matrix - with centroid replacement. Twenty groups were isolated, ranging in
size from three containing only one district each to one with 183 constituent
districts. Further in the grouping procedure, nineteen of these groups were
replaced by three large groups - as indicated in Fig. 10. (Note that squared
distances were used in the grouping). A table was then constructed showing
the mean rank position for each of the twenty groups (ranks are from 1-low to
611-high). From this (Table 5) the three main groups can be separately identified as:

I   High on socio-economic status and low on ethnic status - 'middle-class
     suburbia'.

II   High on family status; low on socio-economic status; and high on ethnic
     status - low income suburbs with considerable proportions of non-British
     immigrants.

III Low on family status; high on ethnic status; low on socio-economic status
     - 'inner city immigrant ghettos'.

Not every sub-group is truly representative of the larger group in which
it is amalgamated; in several cases, a sub-group deviates from the group mean
on one of the three variables, and these can usually be identified clearly
in the linkage tree.

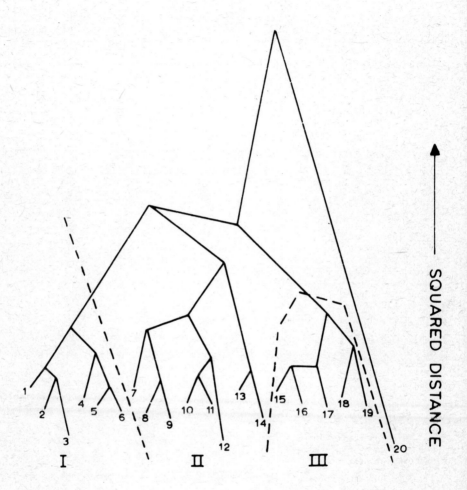

Fig. 10 Linkage tree for the Final Stages of a Classification of
Melbourne's Social Areas (after Jones, 1969, p. 98).

25

Table 5 Mean ranks for Melbourne social areas

(N = 611)

Group	No. of Members	Mean Rank for Socio-Economic Status	Family Status	Ethnic Status
I	287	421	383	155
1	183	375	450	158
2	14	567	534	61
3	1	598	435	1
4	81	519	225	176
5	4	244	209	119
6	3	301	330	15
II	83	104	502	390
7	8	37	602	320
8	41	152	494	407
9	3	226	601	342
10	14	30	463	545
11	5	8	483	607
12	1	36	606	563
13	6	128	518	18
14	5	37	437	149
III	240	236	145	456
15	7	493	7	370
16	47	477	77	402
17	27	265	32	503
18	94	101	144	537
19	66	215	254	362
20	1	57	2	606

Source: Jones (1969, p.99)

III GROUPS AND REGIONS

Our discussion so far has been of the general problems and processes of classification, covering but a small part of the field of numerical taxonomy (Sokal and Sneath, 1963). In some areas of study, geographers are faced with a problem which, although not peculiar to geography is typical of our discipline; this is the question of the analogy between groups and regions (Grigg, 1965, 1967).

The region lies at the heart of much geographical work. Two types of region are usually recognised: (1) formal, comprising places with similar characteristics; and (2) functional, or nodal, comprising places with similar linkage patterns to other places. Each of these can be further subdivided, into (1) regional types, and (2) regions. The difference between the latter two is that whereas the regional type comprises places which are similar on certain predetermined characteristics (landscapes, population structures, etc.), the region also involves a contiguity constraint - a region must comprise spatially conterminous units. Thus we have a 2 x 2 classification scheme, as shown below with an example for each cell:

	Regional Type	Contiguous Region
Formal	Urban Ghettos (e.g. Morrill, 1965)	Political constituencies (e.g. Taylor, 1973)
Functional	World trade groups (e.g. Russett, 1967)	School catchment zones (e.g. Shepherd and Jenkins, 1972)

The production of regional types involves no special procedures; any classification method applicable to the particular data set will be relevant. For contiguous regions, however, special methods may be needed. Among a number that have been suggested (see, for example, Taylor, 1969), two basic approaches can be identified. The first (Czyz, 1968; Johnston, 1970) suggests that the usual classification procedures should be adopted, and that when regional types have been identified, tests should then be made to see if they also form contiguous regions. The alternative, and most frequently used, introduces a contiguity constraint to the grouping procedure.

(i) Contiguity Constraints

The use of contiguity constraints can be illustrated by returning to our sixteen villages. The original map (Fig. 1) shows the location of these, and from it we can derive a contiguity matrix (Table 6), in which a 1 in a cell indicates that the relevant villages (the row and column numbers) have a boundary in common, and an 0 indicates that the two villages are not contiguous. (Note that we may have problems where four boundaries meet; we have decided, for example, that villages 3 and 9, and also 6 and 10, do not have common boundaries).

We now proceed through our nierarchical clustering with centroid replacement procedure as before, but with one difference; before grouping two villages, we check in the matrix to see if they are contiguous and if they are not, then we don't group them and proceed to the next shortest distance. Thus the shortest distance in Table 1 is 3.0, between villages 2 and 3, but Table 6 and Figure 1 show us that these have no common boundary and so would not form a region. The next shortest is 5.4, for villages 3 and 4, but again these are not contiguous. In fact, the shortest distance at which we proceed to a grouping is 7.1, at which villages 2 and 4 are classed together, as are 13, 14 and 15 (but not 11 and 12).

After each grouping we must recompute both the distance and the contiguity matrices, although as an alternative we may compute a matrix in which distances

Table 6 Contiguity Matrix for Figure 2[a]

							Place									
	1	2	3	4	5	6	7	8	9	10	11	12	13	14	15	16
1	-	0	1	1	0	1	0	0	0	0	0	0	0	0	0	0
2	0	-	1	1	0	0	0	0	0	0	0	0	0	0	0	1
3	1	1	-	1	0	1	0	0	0	1	0	1	0	0	0	0
4	1	1	1	-	1	1	0	0	0	0	0	0	0	0	0	0
5	0	0	0	1	-	1	0	1	1	0	0	0	0	0	0	0
6	1	1	1	1	1	-	0	0	1	0	0	0	0	0	0	0
7	0	0	0	0	0	0	-	1	0	0	1	0	1	1	0	1
8	0	0	0	0	1	0	1	-	1	0	1	0	1	1	0	0
9	0	0	0	0	1	1	0	1	-	1	0	0	0	0	0	1
10	0	0	1	0	0	0	0	0	1	-	0	1	0	1	0	0
11	0	0	0	0	0	0	1	1	0	0	-	0	1	0	0	1
12	0	0	1	0	0	0	0	0	0	1	0	-	0	0	1	0
13	0	0	0	0	0	0	1	1	0	0	1	0	-	1	0	0
14	0	0	0	0	0	0	1	1	0	1	0	0	1	-	1	0
15	0	1	0	0	0	0	0	0	0	0	0	1	0	1	-	0
16	0	0	0	0	1	0	1	0	1	0	1	0	0	0	0	-

A_1 = two places are contiguous   0 = non-contiguous

to contiguous units only are entered, and all other distances are replaced by a 'nonsense' value (say 1000.0). The latter, of course, can not be used in computing the new distance matrix; reference back to the original will be necessary.

For our sixteen villages, the full grouping with contiguity constraint proceeded as follows:

Step	Places Grouped	Distance	Forming Group	Average Intra-Group Distance
1	(13-14-15	7.1	A	8.1
	(2-4	7.1	B	7.1
2	B-1	8.6	C	8.1
3	C-3	5.9	D	7.0
4	5-6	10.0	E	10.0
5	(7-8	14.1	F	14.1
	(9-10	14.1	G	14.1
6	E-F	19.6	H	17.1
7	G-12	29.2	J	24.2
8	H-11	33.9	K	23.8
9	D-16	35.9	L	18.6
10	J-K	36.7	M	
11	L-M	50.9	N	
12	A-N	59.2	O	

The linkage tree is given in Figure 11. Comparison between this and that for the unconstrained grouping (Figs. 5 and 11; the villages are placed in the same position on the two horizontal axes to aid comparison) indicates the major differences. These involve first, villages 11 and 12 which, because of their great spatial distance are only grouped together at the last step but two in the constrained solution, as against step three in the unconstrained; and, secondly, the separate identity until the last step of the group of villages closest to the town (13, 14, 15). Note also the interesting point that the distance involved at the third step was shorter than that at the other two, a consequence of the prior grouping of villages 1, 2 and 4; only 1 is adjacent to 3, which is closest in the variable space (Fig. 2) to 2 and 4.

(ii) Grouping and Regionalising: An Example

An example of regionalisation procedure, using Ward's method, is a recent paper by Byfuglien and Nordgard (1974) on farming types in eastern Norway. Their data referred to 86 spatial units (communes and combinations of communes), for each of which they had information on six variables referring to farming activity. The six variables were reduced to four, which were orthogonal, using principal components analysis.

29

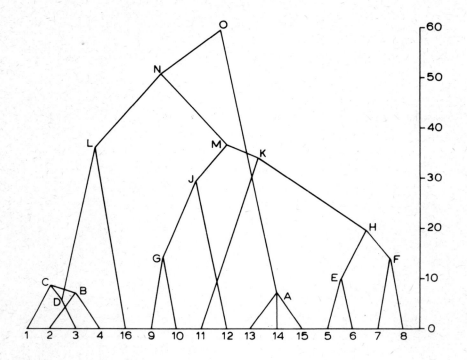

Fig.11   Linkage tree for Hierarchical Clustering with Centroid Replacement
         and Contiguity Constraint of villages in Figure 2.

    In their analyses of these data, the authors produced both a type of
farming areas classification and a farming-types regionalisation. For the former, no contiguity constraint was imposed, and they identified eight different
types of farming areas. Because most of those types comprised several groups
of non-adjacent areas, the resulting map consisted of thirty separate areas
(Figure 12A). In their regionalisation, which used the same procedures plus
a contiguity constraint, sixteen farming-type regions were identified (Figure
12B).

    Comparison of the two maps in Figure 12 will indicate the differences be-
tween the regional type analysis (12A) and the contiguous regions analysis
(12B) in their division of the same area. Clearly, the researcher must be sure
what sort of classification he wishes to produce. As a general guide, it is
probably the case in almost all studies that, unless a contiguous regionalisation is required, the regional type procedure is more 'efficient'. This

point is indicated in Figure 13, where the sum of the squared distances within groups is expressed as a percentage of the sum of all squared distances in the original matrix. At the left of this diagram, where all 86 units are separately identified, the within/total percentage is zero. As one moves to the right, with fewer groups, so the percentage increases, slowly at first and then rapidly. At each step, the regional type analysis percentage is lower than that for the contiguous region analysis, because the former groups together the most similar units but the latter only does so if they are also contiguous.

In this, as in so many other aspects of the classification procedure, choice on the part of the researcher can have a significant influence on the outcome. Contiguous regions are a special case of regional types, and the analyst must be sure, before he proceeds, just what end-product he desires, and what by-products this will produce.

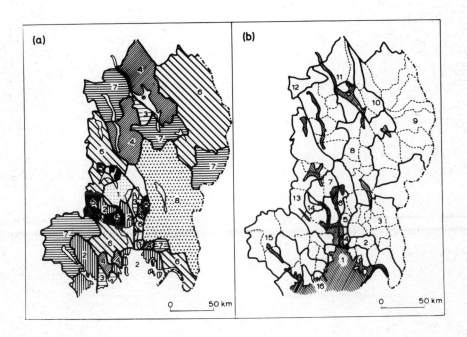

Fig. 12 A division of eastern Norway using Ward's method, A without contiguity constraint, and B with contiguity constraint (after Byfuglien and Nordgard, 1974).

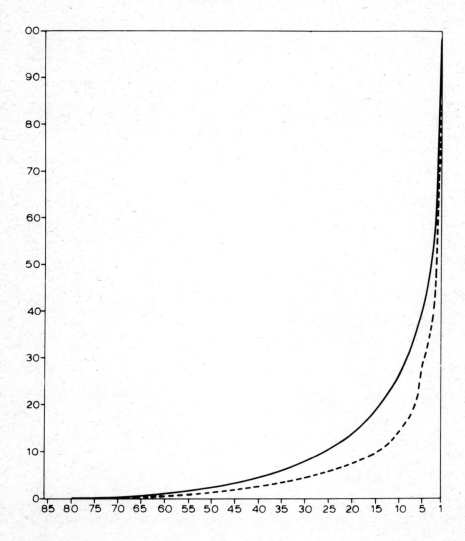

Fig. 13 The efficiency of the classifications shown in Figure 12. On the vertical axis, the within-group distances as a percentage of the total distances are shown; on the horizontal axis the number of groups is indicated. The solid graph refers to the grouping with contiguity constraint, the pecked line to the grouping without such a constraint (after Byfuglien and Nordgard, 1974).

IV DIVISIVE METHODS

Most of the discussion in this brief introduction to classification procedures in geography has been concerned with agglomerative methods, since they
are the most widely used. In them, we start by treating each individual observation as a separate group, and then proceed to group the groups. Another
set of techniques is available, however, which operates in exactly the opposite
way: it begins by assuming that all of the individuals are in one group, and
then proceeds to divide the group into sub-groups. Again, a variety of techniques has been devised - such as Iterative Intercolumnar Correlational Analysis (McQuitty and Clark, 1968) - but we will deal here with only one method,
Association Analysis, which has been fairly widely used for a particular type
of data.

(i) Association Analysis

This method was developed by two botanists, Williams and Lambert (1959),
to cope with the classification of plant communities. It is particularly suit-
ed to nominal data (often known as binary data), which record only the presence or absence of a species (variable) at a site (observation), rather than
any intensity of species' presence. Any data measured at other scales - ordinal,
interval, or ratio - can, of course, be re-written in binary form to make use
of the method.

Association analysis is based on differences between species in their
distributions over sites, and uses the chi-square statistic. An example of
its operation is given here using the hypothetical data of Table 7, which
shows the presence/absence of the nests of eight different bird species (A--
-H) in twenty different woodland samples. Presence is shown by a 1 in the
relevant cell; absence by a zero. (We have taken presence/absence as the divid-
ing line, but we could have coded, say, any woodland with less than five nests
per 100 metre square as zero).

The grouping proceeds as follows. First, the chi-square (χ^2) statistic
for the distributions of each pair of species is calculated. This involves
preparation of a 2 x 2 table:

		Species x		
		present	absent	
Species y	present	a	b	a+b
	absent	c	d	c+d
		a+c	b+d	N

Cell a lists the number of sites (woodlands) in which both of the species are
present, call b those sites containing species y but not species x, and so on:
N is the total number of sites. Chi-square is then computed, using the common
formula for its application to a 2 x 2 table:

$$\chi^2 = \frac{(ad-bc)^2 N}{(a+b)(a+c)(b+d)(c+d)} \quad \dots\dots\dots\dots\dots (6)$$

Table 7 Distribution of bird species by woodlands

Woodland	Species							
	A	B	C	D	E	F	G	H
1	1	0	1	0	1	0	1	0
2	1	1	1	0	0	0	1	1
3	1	1	1	0	0	0	0	0
4	1	1	1	0	0	0	1	1
5	0	0	0	1	1	1	0	0
6	1	0	0	0	1	1	1	1
7	0	0	1	0	0	0	0	0
8	1	1	1	0	0	0	1	1
9	0	1	1	1	0	0	1	0
10	1	1	1	0	0	0	0	1
11	0	0	1	1	0	0	1	0
12	1	0	1	0	0	0	1	1
13	0	1	0	1	1	1	0	0
14	0	0	0	0	1	1	1	1
15	1	0	0	0	1	0	0	0
16	0	0	0	1	0	1	1	1
17	0	1	0	1	1	1	0	1
18	0	0	0	1	1	0	1	0
19	0	0	0	1	1	1	0	1
20	0	0	0	1	1	1	1	0

For species A and B in Table 7, the 2 x 2 contingency table and the calculation for χ^2 are:

Species A

Species B		present	absent	
	present	5	3	8
	absent	4	8	12
		9	11	20

$$\chi^2 = \frac{((5\text{x}8) - (4\text{x}3))^2 \text{ x } 20}{(5+3) \text{ x } (5+4) \text{ x } (3+8) \text{ x } (4+8)}$$

$$= ((40-12)^2 \text{ x } 20)/8 \text{ x } 9 \text{ x } 11 \text{ x } 12 = (28^2 \text{ x } 20)/9504$$

$$= (784 \times 20)/9504 = 15680/9504 = 1.6$$

Table 8 Inter-Species chi-square statistics

				Species				
	A	B	C	D	E	F	G	H
A	-	1.6	5.1	13.4	1.8	5.7	0.4	1.8
B	1.6	-	3.3	0.4	3.0	1.2	0.5	0.8
C	5.1	3.3	-	5.1	12.8	13.3	0.8	0
D	13.4	0.4	5.1	-	1.8	5.1	0.1	1.8
E	1.8	3.0	12.8	1.8	-	7.5	0.8	0.8
F	5.7	1.2	13.3	5.1	7.5	-	0.6	0.8
G	0.4	0.5	0.8	0.1	0.8	0.6	-	0.8
H	1.8	0.8	0	1.8	0.8	0.8	0.8	-
Σ	29.8	10.8	40.4	27.7	28.5	34.2	4.0	6.8

The full 8 x 8 matrix of inter-species χ^2 values is presented in Table 8.

At the next stage of the procedure, we decide which species to use as the one to divide the woodlands into two groups. The species chosen is that which is most dissimilar from all of the others, and therefore is assumed to be the best discriminator between types of woodlands. Since the size of an individual χ^2 is positively related to the degree of dissimilarity between two species' distributions, then the most dissimilar overall is that species with the largest sum of its seven χ^2 values. So the column sums are computed, and are given in Table 8. From these, it is seen that species C is the most dissimilar and this forms the basis of the division: all woodlands containing species C are placed in one group, and those without species C in another, giving the following group membership:

Group	Members
I - C present	1,2,3,4,7,8,9,10,11,12
II - C absent	5,6,13,14,15,16,17,18,19,20

We now have two groups and the full procedure is repeated separately for each. Distribution tables for the seven remaining species are drawn up (Table 9); species C is no further use in the classification, since within either group there is no variance - it is either always present or always absent. Two separate 7 x 7 matrices of χ^2 values are then derived (Table 10); the resulting sums indicate that group I should be divided on the basis of presence or absence of species A, whereas species D is the discriminator within group II. This gives four groups, on each of which the process is repeated.

Division continues until all groups contain two or fewer members, beyond which further classification would be meaningless. Occasionally larger groups

35

Table 9 Distribution of bird species by two woodland groups

	Group I								Group II						
	Species								Species						
Woodland	A	B	D	E	F	G	H	Woodland	A	B	D	E	F	G	H
1	1	0	0	1	0	1	0	5	0	0	1	1	1	0	0
2	1	1	0	0	0	1	1	6	1	0	0	1	1	1	1
3	1	1	0	0	0	0	0	13	0	1	1	1	1	0	0
4	1	1	0	0	0	1	1	14	0	0	0	1	1	1	1
7	0	0	0	0	0	0	0	15	1	0	0	1	0	0	0
8	1	1	0	0	0	1	1	16	0	0	1	0	1	1	1
9	0	1	1	0	0	1	0	17	0	1	1	1	1	0	1
10	1	1	0	0	0	0	1	18	0	0	1	1	0	1	0
11	0	0	1	0	0	1	0	19	0	0	1	1	1	0	1
12	1	0	0	0	0	1	1	20	0	0	1	1	1	1	0

Table 10 Inter-species chi-square statistics

	Group I							Group II						
	Species							Species						
	A	B	D	E	F	G	H	A	B	D	E	F	G	H
A	-	1.3	6.0	0.5	0	0	4.3	-	0.6	6.0	0.3	1.4	0	0
B	1.3	-	0.1	1.6	0	0.1	1.6	0.6	-	1.1	0.3	0.6	2.5	0
D	6.0	0.1	-	0.3	0	1.1	2.5	6.0	1.1	-	0.5	0.5	0.5	0.5
E	0.5	1.6	0.3	-	0	0.5	1.1	0.3	0.3	0.5	-	0.3	1.1	1.1
F	0	0	0	0	-	0	0	1.4	0.6	0.5	0.3	-	0	2.5
G	0	0.1	1.1	0.5	0	-	0.5	0	2.5	0.5	1.1	0	-	0.4
H	4.3	1.6	2.5	1.1	0	0.5	-	0	0	0.5	1.1	2.5	0.4	-
Σ	12.1	4.7	10.0	4.0	0	2.2	10.0	8.3	6.1	9.1	3.6	5.3	4.5	4.5

can not be divided further, because the members are all completely alike. In our example, woodlands 2, 4, and 8 form a group of 3 at the fourth step. These three each have a presence score for three of the species still being considered (B,G and H) and an absence score for the other two (D and F); they are alike in all respects, and the only further division is into unique individual woodlands. (Their uniqueness would not be on the criteria of the eight bird species being used in the classification, but some other criterion; location perhaps). The full grouping proceeds as follows:

Step	Group	Divide on Species		Forms Groups of Woodlands	Largest χ^2
1		C	I	1,2,3,4,7,8,9,10,11,12	13.3
			II	5,6,13,14,15,16,17,18,19,20	
2	I	A	III	1,2,3,4,8,10,12.IV 7,9,11	6.0
	II	D	V	5,13,16,17,18,19,20.VI 6,14,15	6.0
3	III	E	VII	1 XIII 2,3,4,8,10,12	2.9
	IV	D and G	IX	7 X 9,11	3.0
	V	G	XI	5,13,17,19 XII 16,18,20	2.1
	VI	F,G, and H	XIII	15 XIV 6,14	3.0
4	VIII	G	XV	2,4,8,12 XVI 3,10	2.8
	XI	B and H		four groups of one member each	
	XII	E and H	XVII	16 XVIII 18,20	3.0
5	XV	B	XIX	2,4,8 XX 12	-
6	XIX			division not possible - three members perfectly alike	

As with the agglomerative procedures, we are faced with the question 'when should the grouping be stopped?'. The answer is generally arbitrary, though a frequently-applied rule is when the maximum value of χ^2 for the species on which the division is made is less than 3.84 (which is the minimum value of χ^2 to show a significant difference at the .05 level of probability when there is only one degree of freedom; Gregory, 1972). Choice of the .05 level rather than any other reflects convention. Furthermore, statistical significance testing is irrelevant where samples are not being used. The rule provides a useful assessment device, however.

Maximum χ^2 values are given in the above summary of the grouping procedure, indicating that - on the statistical significance criterion - the process should end after step 2. A linkage tree with the χ^2 values on the vertical axis indicates the procedure (Fig. 14). In this the letters at the breaks in the divisive process indicate the species on which the division was made, the number of members of each group is indicated within the circle, the group identification is shown by Roman numerals and the positive (presence) and negative (absence) signs indicate which group has the species on which the split takes place, and which has not.

37

As with all of the other techniques discussed here, a range of minor variations on the association analysis theme has been introduced, using different coefficients to χ^2, different stopping rules, and so on. No contiguity constraint can be built in, however, and the method produces regional types only. One advantage of the method is that it is easily applied to the data matrix in both dimensions. Our analysis analysed differences between the variables (the bird species) and classified the woodlands, but we could just as easily have calculated χ^2 values between each pair of woodlands and so classified the bird species; the former method is known as 'normal' association analysis and the latter as 'inverse' association analysis.

Association analysis has been widely used in ecological and biogeographical work, whose data are frequently in binary form (Frenkel and Harrison, 1974), but it is of value in a range of other fields. Caroe (1968), for example, used both the 'normal' and the 'inverse' approach in a study of the functional structure of East Anglian settlements in 1846. The data comprised the presence/absence coding of 61 trade and service functions for 76 settlements. Normal analysis suggested eleven groups of settlements at the .05 level, whereas inverse analysis indicated twelve groups of central functions. Combination of these two groupings into a diagram (Fig. 15), in which a dot represents the presence of a function in that settlement, suggests a clear hierarchical central place system ranging from the regional market centres through to the small villages. This is, in effect, a re-ordering of the original data matrix, in a form which highlights the order among its 4636 cells.

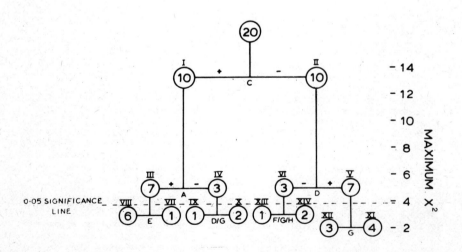

Fig. 14 Linkage tree for Association Analysis of the Woodlands in Table 7.

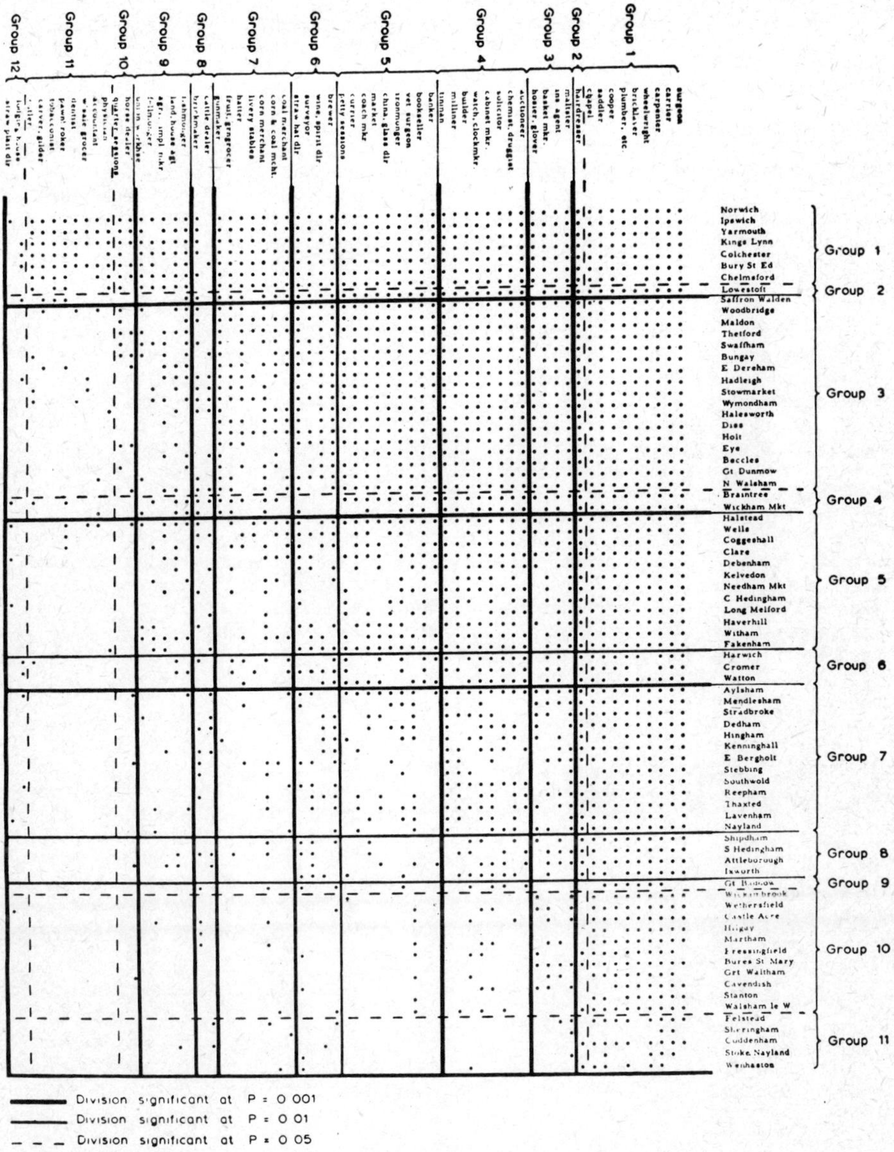

Group 12
Group 11
Group 10
Group 9
Group 8
Group 7
Group 6
Group 5
Group 4
Group 3
Group 2
Group 1

surgeon
carrier
carpenter
wheelwright
plumber, etc.
cooper
saddler
vet agent
ins agent
basket maker
hatter glover
cheesemonger/druggist
solicitor
cabinet mkr
watch, clockmkr
builder
milliner
tinman
bookseller
vet surgeon
china, glass dir
market
coach mkr
corn merchant
corn & coal mchl.
miller
area hat dir
surveyor
wine, spirit dir
brewer
straw bonnet dir
coal merchant
cattle dealer
brickmaker
lime burner
agr, impl dir
agr, horse agt
dentist
wool grocer
basket maker
tobacconist
straw plait dir
carver, gilder

Norwich
Ipswich
Yarmouth
Kings Lynn
Colchester
Bury St Ed
Chelmsford — Group 1
Lowestoft — Group 2
Saffron Walden
Woodbridge
Maldon
Thetford
Swaffham
Bungay
E Dereham
Hadleigh
Stowmarket
Wymondham — Group 3
Halesworth
Diss
Holt
Eye
Beccles
Gt Dunmow
N Walsham
Braintree — Group 4
Wickham Mkt
Halstead
Wells
Coggeshall
Clare
Debenham
Keivedon
Needham Mkt — Group 5
C Hedingham
Long Melford
Haverhill
Witham
Fakenham
Harwich
Cromer — Group 6
Watton
Aylsham
Mendlesham
Stradbroke
Dedham
Hingham
Kenninghall
E Bergholt — Group 7
Stebbing
Southwold
Reepham
Thaxted
Lavenham
Nayland
Shipdham
S Hedingham — Group 8
Attleborough
Ixworth
Gt Bardow — Group 9
Wickhambrook
Wethersfield
Castle Acre
Hingay
Martham
Fressingfield
Bures St Mary — Group 10
Gt Waltham
Cavendish
Stanton
Walsham le W
Felstead
Sheringham
Coddenham — Group 11
Stoke Nayland
Wenhaston

Division significant at P = O 001
Division significant at P = O 01
Division significant at P = O 05

ASSOCIATION ANALYSIS: FINAL GROUPING BY TRADES AND SERVICES
AND BY SETTLEMENTS, 1846

Fig. 15 Groups of Settlements and of Central Functions in East Anglia
1876, according to Association Analysis (after Caroe, 1968, p.264).

39

V CONCLUSIONS

Classification is a large field, as a number of recent reviews suggest (Everitt, 1974; Cormack, 1971; Sibson and Jardine, 1971). Only a relatively small proportion of the available methods has been used in geographical research, but even so, this means considerable variation within the literature in terms of actual procedures employed. Only some of the simpler, more easily applied techniques have been discussed here, though the rules on which they operate are similar to those of many others which space did not allow us to consider.

Perhaps three main related conclusions can be drawn from the examples we have discussed. The first is that, although the procedures we have investigated are fairly straightforward, if we are dealing with a data set of any magnitude - particularly in the number of observations to be grouped (as in Jones' study) - then the process rapidly becomes tedious and time-consuming, and therefore probably prone to simple error. Fortunately, computer programs are now widely available to remove the tedium, but herein lies the second conclusion. Careless choice of method may produce results which could well be misleading, and one of the dangers of the availability of programs - for many statistical routines and not just classification - is that people may choose a method simply because a program is available rather than on logical grounds. Where a program offers a large number of options (such as the popular CLUSTAN routine: Wishart, 1969), choice may be even more haphazard. Another problem with computer routines is that most of them, because of limited memory size, cannot cope with large data sets and distance matrices, leading several researchers (e.g. Pocock and Wishart, 1969) to investigate possible ways of reducing the labour.

The final conclusion stems from the first two. Although we realise the necessity for classification, there has been little discussion among geographers about the definitional problems involved. We do not have well-developed theories of regions and of regional types to compare with those developed for their own purposes by, say, educational psychologists (McQuitty, 1967). Grigg (1965, 1967) initiated some discussion on the basis of his careful outline of the logic of regionalisation (see Bunge, 1966a; Grigg, 1966), but most geographers have been concerned more with finding a technique and getting on with the classification rather than with making sure for what it is they are classifying.

Classification is a crucial stage in scientific development, therefore. All of the technical problems have not been solved (despite Bunge's 1966b claim), but a range of easily-applied methods is available to the researcher. Selection among these requires some choice, and therefore careful consideration of the purpose of the study, as does evaluation of the output.

BIBLIOGRAPHY

1. Berry, B.J.L. and P.H. Rees (1969) The factorial ecology of Calcutta , *American Journal of Sociology*, 74, 445-491.

2. Bunge, W. (1966a) Locations are not unique , *Annals, Association of American Geographers*, 56, 376-377.

3. Bunge, W. (1966b) 'Gerrymandering, geography and grouping , *Geographical Review*, 56, 256-263.

4. Byfuglien, J. and A. Nordgard (1974) Types or regions? , *Norsk Geografisk Tidsskrift*, 28, 157-166.

5. Caroe, L. (1968) A multivariate grouping scheme: association analysis of East Anglian towns , in E.G. Bowen, H. Carter and J.A. Taylor (eds) *Geography at Aberystwyth* (University of Wales, Cardiff), 253-269.

6. Cormack, R.N. (1971) A review of classification , *Journal, Royal Statistical Society*, A, 134, 321-367.

7. Czyz, T. (1968) The application of multifactor analysis in economic regionalization , *Geographia Polonia*, 15, 115-134.

8. Everitt, B. (1974) *Cluster Analysis* (Heinemann, London).

9. Frenkel, R.E. and C.M. Harrison (1974) An assessment of the usefulness of phytosociological and numerical classificatory methods for the community biogeographer , *Journal of Biogeography*, 1, 27-56.

10. Gregory, S. (1972) *Statistical Methods and the Geographer*, (Longmans, London).

11. Grigg, D.B. (1965) The logic of regional systems , *Annals, Association of American Geographers*, 55, 465-491.

12. Grigg, D.B. (1966) Are Locations Unique? A Reply to W. Bunge , *Annals, Association of American Geographers*, 56, 376-377.

13. Grigg, D.B. (1967) Regions, Models and Classes , in R.J. Chorley and P. Haggett (eds) *Models in Geography*, (Methuen, London) 461-509.

14. Johnston, R.J. (1965) Multivariate regions - a further approach , *The Professional Geographer*, 17, 9-12.

15. Johnston, R.J. (1968) Choice in classification: the subjectivity of objective methods , *Annals, Association of American Geographers*, 58, 575-589.

16. Johnston, R.J. (1970) Grouping and regionalising: some methodological and technical observations , *Economic Geography*, 46, 293-305.

17. Johnston, R.J. (1971) *Urban Residential Patterns*, (G. Bell and Sons, London).

18. Jones, F.L. (1969) *Dimensions of Urban Social Structure*, (A.N.U. Press, Canberra).

19. King, L.J. (1969) *Statistical Analysis in Geography*, (Prentice-Hall, Englewood Cliffs).

20. Lance, G.N. and W.T. Williams (1967) 'A general theory of classificatory sorting strategies , *Computer Journal*, 9, 373-380.

21. McQuitty, L.L. (1967) A mutual development of some typological theories and pattern-analytic methods , *Educational and Psychological Measurement*, 27, 21-46.

22. McQuitty, L.L. and J.A. Clark Clusters from iterative, intercolumnar correlation analysis , *Educational and Psychological Measurement*, 28, 211-238.

23. Morrill R.L. (1965) The negro ghetto , *Geographical Review*, 55, 339-361.

24. Pocock, D.C.D. and D. Wishart (1969) Methods of deriving multi-factor uniform regions , *Transactions, Institute of British Geographers*, 47, 73-98.

25. Russett, B.M. (1967) *International Regions and the International System*, (Rand McNally, Chicago).

26. Shepherd, J.W. and M.A. Jenkins (1972) Decentralizing highschool administration in Detroit: an evaluation of alternative strategies of political control , *Economic Geography*, 48, 95-106.

27. Sibson, R. and N. Jardine (1971) *Mathematical Taxonomy*, (J. Wiley, London).

28. Smith, R.H.T. (1966) Method and purpose in functional town classification , *Annals, Association of American Geographers*, 56, 539-548.

29. Sokal, R.R. and P.H.A. Sneath (1963) *Principles of Numerical Taxonomy*, (W.H. Freeman, San Francisco).

30. Spence, N.A. and P.J. Taylor (1970) Quantitative methods in regional taxonomy , *Progress in Geography*, 2, 1-64.

31. Taylor, P.J. (1969) The location variable in taxonomy , *Geographical Analysis*, 1, 181-195.

32. Taylor, P.J. (1973) Some implications of the spatial organization of elections , *Transactions, Institute of British Geographers,* 60, 121-136.

33. Ward, J.H. Jr (1963) Hierarchical grouping to optimise an objective function , *Journal, American Statistical Association,* 58, 236-244.

34. Williams, W.T. and J.M. Lambert (1959) Multivariate methods in plant ecology. 1. Association analysis in plant communities , *Journal of Ecology,* 47, 83-101.

35. Wishart, D. (1969) Fortran II programs for 8 methods of cluster analysis (CLUSTAN II), *Computer Contribution 39,* State Geological Survey, University of Kansas.